Bromley
THE LONDON BOROUGH

SHORTLANDS LIBRARY
020 8460 9692

Please return this item by the last date stamped below, to the library from which it was borrowed.

Renewals
Any item may be renewed twice by telephone or post, provided it is not required by another customer. Please quote the barcode number.

Overdue Charges
Please see library notices for the current rate of charges levied on overdue items. Please note that the overdue charges are made on junior books borrowed on adult tickets.

Postage
Postage on overdue notices is payable.

2 3 MAR 2005	1 SEP	- 9 MAY 2008
1 2 JUL 2005	22 SEP	1 2 SEP 2008
		3 0 JAN 2009
BEC	1 3 OCT 2007	BEC
1 9 JAN 2007	3 NOV 15 DEC	27 12
1 7 MAR 2007	BEC	1 1 FEB 2011
- 4 APR 2007		- 8 AUG 2011
1 1 JUN 2007	1 9 JAN 2008	1 8 FEB 2012
30 JUN	- 7 MAR 2008	1 2 NOV 2013
1 1 AUG 2007	28/3/08	1 9 JAN 2024
	1 5 APR 2008	

92.04

THE WHIC

WIRI
LIGH